MATHEMATICS

Computations

Steps To High School Math

Volume I

CSEC/GSCE

Grade 7-11

Printed & Fulfilled by:
Printbindship by Mel
and Amazon KDP.

ISBN:
9798870359090
Printed in the U.S.A

Contact sales: sales@dbookship.com

INTRODUCTION

This book is a brief exploration of the GSCE and CSEC syllabus. The purpose of this book, is to prepare students for exam conditions, where they will be able to handle just about any questions thrown regarding Consumer Arithmetics. This book outlines the basic computation problems and reasonings behind Consumer Arithmetics, especially as it relates to currency and buying and selling. A lot of fun ideas, shortcuts and tricks were integrated into this book for students to perform various calculations at the highest level with minimal chances of making any mistakes. Please take the time out to go through each topics and do the practice activities given at the end of each chapter of this book.

<u>ACKNOWLEDGEMENT</u>

In this note, I take privilege to use this opportunity to thank my supporters on Tiktok, Facebook and Instagram for the kind support and motivation you have given me. I started this journey without knowing that I would become an author to publish Mathematical contents to inspire and strengthen our future. It was through your inspiring words and your support of my work, that I received the idea to make blissful content for you. My desire is that I will be able to reach thousands more people who are having struggles mathematically, that I may be able to shape this world more effectively. Please continue to push me and support me on this journey, as I will be looking to be publishing more learning contents to continue the great work of revolutionizing our future.

TABLE OF CONTENT

Content   Pages

FRACTIONS

Fractions: Simply put, fraction is a portion of a whole. A fraction is therefore a value less than one (1). Every fraction can be expressed in decimal form. As we progress through this lesson you will see that fraction is simply an expression of division in another way.

In a fraction, the number above is the 'numerator' and the bottom number is the 'denominator'. When converting to decimals, we say that the numerator becomes the 'dividend' and the denominator is the 'divisor'. The result of division is called the quotient!

$$\frac{1}{4} \xrightarrow{\text{Divisor}} 4 \overline{)1\,00} \quad \begin{array}{r} 0.25 \leftarrow \text{Quotient} \\ \hline 1\,00 \leftarrow \text{Dividend} \\ -\,8 \\ \hline 20 \\ -\,20 \\ \hline \cdot\,\cdot \end{array}$$

The numerator **1** becomes the dividend. We initiate our decimal figures by adding **0** to the dividend until there are no more remainders.

With this understanding, we can say this fraction is an expression of $1 \div 4$. Another way to express division is as follows:

$$\frac{\frac{1}{4}}{6} \;=>\; \frac{1}{4} \;\div\; \frac{1}{6}$$

Fractions are very unique when performing various operations. To add or subtract fractions, the denominators MUST be the same. If the denominators are different, we go by finding the 'Lowest Common Multiple (L.C.M)'.

<u>Consider the following</u>:

$$\frac{2}{5} + \frac{1}{5} = \frac{3}{5} \quad \begin{array}{l} (\text{ Add the numerator}) \\ (\text{ We simply re-write the denominator}) \end{array}$$

1

The decimal value of 2/5 + 1/2 literally equals the decimal value of 3/5. This makes the equation true (0.4) + 0.2 = 0.6). However, if we are to try the same operations with fractions of different denominator, the result would be different.

Consider the following:

$$\frac{2}{5} + \frac{1}{4} =$$

a). $\frac{3}{5}$

b). $\frac{3}{4}$

c). $\frac{3}{9}$

d). None of the above

Let us test our equation to see which of the 3 fractions above would make our equation true. We achieve this by checking the decimal values of each fraction:

2/5 = 0.4 + 1/4 = 0.25 => **0.65**
3/5 = **0.6**
3/4 = **0.75**
3/9 = **0.333...**

We can see here that none of our solutions would make the equation evaluates true! So our answer would be **(d) None of the above**.
So what then is our answer? How then do we solve this mathematical problem for our evaluation to be true everytime?

- The rule is, only fractions with the same denomainator can be added or subtracted. Rewrite the denominator and add the enumerators.
- However, if we have different denominator then we find a common denominator using the lowest common multiple amongst each pair of fractions.

Consider the following:

$$\frac{2}{(5)} + \frac{1}{(4)}$$

LCM = 20

| 5 (multiples of 5) => 5,10, 15, 20, 25, 30, 35, 40...
| 4 (multiples of 4) => 4, 8,12, 16, 20, 24, 28, 32, 36, 40 ...
|
| Here we see two common multiples 20 & 40. There are more common multiples between 5 and 4, that we could use have used. Choosing any of the C.M would give us the same results. However, choosing the L.C.M makes our mathematical calculations simpler or less complex.

$$\frac{(2)^{x4}}{(5)} + \frac{(1)^{x5}}{(4)}$$

$$\frac{8 + 5}{20} = \frac{13}{20}$$

| We use our LCM to be our new denominator for both fractions. Then we divide the old fraction into the LCM, quotient is multiplied by the numerator. So our new fractions become 8/20 + 5/20 = 13/20 (Adding the numerator and rewrite the denominator).

Now let us confirm our answer by checking its decimal value:

```
        0.6 5
   20 | 130
        -120
         100
        -100
          . .
```

| We initiate our decimal values by adding 0 to the ones place of 13. Then we proceed with our division 130 ÷ 20 = 0.65.
| We finally have our results and we see that our equation => (2/5 + 1/4 = 13/20) = (0.4 + 0.25 = 0.65) evaluates true!

In our previous exercises we have covered how to convert fractions to decimals and also adding or subtracting fractions. We also said that any common multiples could be used for the denominator.

<u>Let us look at the example below:</u>

$$\frac{2}{5} + \frac{1}{4} \quad \underline{LCM= 20}$$

$$\frac{8}{20} + \frac{5}{20} = \frac{13}{20}$$

$$\frac{2}{5} + \frac{1}{4} \quad \underline{CM=40}$$

$$\frac{16}{40} + \frac{10}{40} = \frac{26}{40}$$

Equivalent Fractions

Both 13/20 and 26/40 equals **0.65**. You may also realize that 26/40 is exactly twice the size of 13/20. Where I am going with this is that I need students to understand how equivalent fractions work. If you look closely, when we find the **LCM** of 5 & 4, the fraction is converted to their equivalent with a denominator set to the **LCM**; 8/20 is 4 times the size of 2/5 and 5/20 is 5 times the size of 1/4. So when you are adding or subtracting fractions with different denominators, we find a pair of equivalent fractions for both terms, where the denominators are the same!

Equivalent fractions have the same decimal values and can be substituted to solve mathematical equations if one equivalent is too complex or difficult to work with. We obtain the equivalent of a fraction as follows:

$$\frac{13}{20} \begin{smallmatrix} x2 \\ x2 \end{smallmatrix} = \frac{26}{40}, \quad \frac{13}{20} \begin{smallmatrix} x3 \\ x3 \end{smallmatrix} = \frac{39}{60}, \quad \frac{13}{20} \begin{smallmatrix} x4 \\ x4 \end{smallmatrix} = \frac{52}{80} \; \dots$$

All the above fractions have the same decimal values!

When we break down a fraction to their lowest terms by dividing both numerator and denominator by a common factor (C.F), we are also finding the fraction's equivalent.

$$\frac{39 \div 3}{60 \div 3} = \frac{13}{20} \quad \text{and} \quad \frac{52 \div 4}{80 \div 4} = \frac{13}{20}$$

Types of Fraction

There are (3) types of fractions:
- Proper Fractions
- Improper Fractions
- Mixed Fractions

Proper Fractions have a numerator smaller than the denominator, where as Improper Fractions have a numerator greater than the denominator. Mixed Fractions consist of a whole number (number on the left of the decimal point), and a fraction. This is basic information, however students face challenges when working with the different types of fractions.

In most cases, Mixed Fractions are converted into Improper Fractions for mathematical computations. The decimal value of the fraction is still not affected in the conversion. We convert a Mixed Fraction to an Improper Fraction by multiplying the whole number by the denominator and add the product to the numerator. Let us look at an example on how this is done.

$$\left(2 \frac{1}{4}\right) - \left(1 \frac{3}{5}\right) \Rightarrow \frac{9}{4}^{x5} - \frac{8}{5}^{x4}$$

$$\frac{45 - 32}{20} = \frac{13}{20}$$

Some persons would argue that converting to Improper Fractions is a waste of time. However, most students would end up in error or confusion if they were to try to solve this equation without converting to Improper Fractions.

Consider the following:

$$(2) \; \overset{x5}{\underset{4}{1}} \; - \; (1) \; \overset{x4}{\underset{5}{3}}$$

$$(1)\; \frac{5 \quad - \quad 12}{20} \; = \; 1 \; \frac{-7}{20}$$

We start by subtracting our whole numbers and then we move to the fractions. We can see clearly where this method have us subtracting a large number from a smaller number, resulting a negative difference. Most students get stuck right here and may not know how to solve the problem.

Let us look at our problem:

$$\overset{x20}{\underset{1}{1}} \; - \; \overset{x1}{\underset{20}{7}}$$

$$\frac{20 \; - \; 7}{20} \; = \; \frac{13}{20}$$

We write 1/1 and find the LCM of both fractions. We divide our denominators into our LCM and the result we times it by our numerator. Then we complete the operation of subtracting our numerators 20 and 7, which equals 13, and rewrite our LCM as our new denominator. So that is how we would achieve our answers, therefore to save you the trouble, I normally just recommend converting to Improper Fractions before solving!

How to easily calculate your LCM:

I have worked out a way to get my LCM with ease when working with two fractions. You may feel free to exploit this method to work out more fractions, however please take note of my little **DISCLAIMER**: "THIS IS INFORMATION IS FOR GENERAL PURPOSE KNOWLEDGE ONLY, NOT TO SUBSTITUTE FOR COMMON CORE TEACHINGS. USE THIS INFORMATION AT YOUR OWN RISK".

I have never seen this done anywhere. I got the idea from doing Math tricks videos. Let us look at 2 examples.

- To find the LCM of 4 and 6

$$(\frac{2}{4}) \times 3 \overset{+}{} (\frac{5}{6}) \times 2 \Rightarrow \frac{6}{12} + \frac{10}{12} = \frac{16}{12}$$

$$\frac{\cancel{4}^{2}}{\cancel{6}_{3}}$$

- To find the LCM of 5 and 7

$$(\frac{3}{5}) \times 7 \overset{+}{} (\frac{5}{7}) \times 5 \Rightarrow \frac{21}{35} + \frac{25}{35} = \frac{46}{35}$$

$$\frac{5}{7}$$ (There are **NO** common factors between 5 and 7)

Notice that the denominators cross when we supply our fractions with a multiplier. This happened in both cases. The (6) broken down to (3) supplies multiplies the other fraction, and so did the (4) broken down to (2), multiplying the second fraction in our fist example. In the second example, because there are no common factors between 5 and 7, we just simply cross multiply to solve the equation.

Multiplication & Division of Fractions:

We divide fractions by reciprocating the second fraction (fraction to the right) and changing the sign from division (÷) to multiplication (x). We may multiply across (numerator **x** numerator and denominator **x** denominator) or we cancel down our figures if there are any common factors. Factors of a number is the direct opposite of multiples of a number. We say what can go into a number (divided) without leaving any remainders.

Some numbers such as 2, 3, 5 and 7 can only be divided by and itself. These are called Prime Numbers. Other numbers that have more than two (2) factors are called Composite Numbers.

Let us take a look at a work out Math problem.

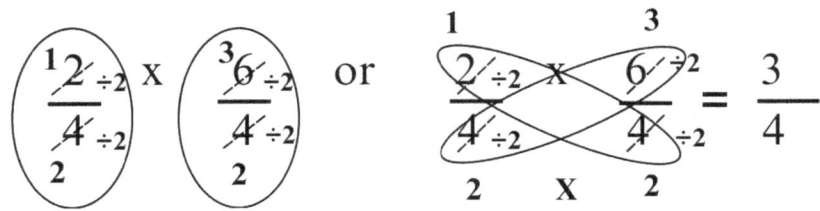

from $\dfrac{2}{4} \div \dfrac{4}{6}$ to ⬭$\dfrac{2}{4}$⬭ x ⬭$\dfrac{6}{4}$⬭ $=$ $\dfrac{\overset{3}{\cancel{12}} \div 4}{\underset{4}{\cancel{16}} \div 4}$

or

$\left(\dfrac{\overset{1}{\cancel{2}} \div 2}{\underset{2}{\cancel{4}} \div 2}\right)$ x $\left(\dfrac{\overset{3}{\cancel{6}} \div 2}{\underset{2}{\cancel{4}} \div 2}\right)$ or $\dfrac{\overset{1}{\cancel{2}} \div 2}{\underset{2}{\cancel{4}} \div 2}$ x $\dfrac{\overset{3}{\cancel{6}} \div 2}{\underset{2}{\cancel{4}} \div 2}$ $=$ $\dfrac{3}{4}$

Either way we cross multiply or cancel down we still achieved our answers. Also notice for multiplication and division, we did not re-write the denominators but multiplied them instead.

Practice Activity

a). $\dfrac{2\frac{2}{3} + 1\frac{5}{6}}{\frac{2}{3}}$

b). $\dfrac{5\frac{1}{4} - 2\frac{1}{3}}{2\frac{1}{2}}$

c). $\dfrac{\frac{1}{5} \text{ x } \frac{3}{10}}{\frac{2}{5}}$

d). $\left(3\frac{2}{7} + 1\frac{2}{3}\right) \div 1\frac{1}{7}$

e). $\dfrac{5\frac{2}{7} + 3\frac{5}{7}}{4 - 2\frac{4}{5}}$

f). $\dfrac{2\frac{1}{3} - 1\frac{1}{2}}{1\frac{5}{6}}$

g). $\dfrac{1}{7} + \left(\dfrac{2}{3} \div \dfrac{11}{12}\right)$

8

DECIMALS

Decimals are smaller units than a whole. We write a decimal value by inserting a decimal point to the write of whole numbers. All decimals are the values of fractions. Therefore decimals can also be converted into fractions.

$$0.25 = \frac{25}{100} \begin{array}{l} \div 25 = 1 \\ \div 25 = 4 \end{array}$$

There are only two (2) digits beyond the decimal point, 2 tenths and 5 hundredths. We can say one of two things here. There are two digits, so I say $10^2 = 100$. I could also say we need to convert up to a hundredths place since that's where our last decimal value is. We then cancel down 25/100 by the HCF (25) and our answer is 1/4.

There are fractions with number values recurring, these are called irregular numbers. Examples of these numbers are: 2/3, 1/3, 22/7 (pi) etc. In Mathematics we may need to represent these values or record the values as data. You may ask the question of how do we record something that doesn't stop repeating or recurring. Mathematicians have develop ways to aid our calculations by using scientific notations, rounding-off to decimal places and significant figures to represent this kind of data.

Scientific Notation/Standard Form

This is done by the powers of 10s. Only 1 single non-zero digit is allowed beyond the decimal point. The value of the original number is still preserved in computation.

6538.83 is expressed as $6.538.83 \times 10^3$ (three decimal places to the left)

We may also have figures such as: 0.0352179 => 3.52179×10^{-2}(Once the number beyond the decimal point is a zero, you can already know our tenth power is going to be a negative, the decimal point will move to its right). Another way to look at it is $3.52179 \div 10^2$ (The sign changes to division and the power is kept positive).

Let us look at our Indices:

Positive

10^6 = 1,000,000 10^5 = 100,000 10^4 = 10,000 10^3 = 1,000 10^2 = 100 and 10^1 = 10.

Negative

10^{-6} = $1/10^6$ 10^{-5} = $1/10^5$ 10^{-4} = $1/10^4$ 10^{-3} = $1/10^3$ 10^{-2} = $1/10^2$ and 10^{-1} = $1/10^1$.

What if you see a number such as 1.567872, what is the scientific notation? Since there is already one non-zero digit on the left of the decimal point we wouldn't have to do anything. We write 1.567872 x 10^0(which equals 1.567872 x 1), thus preserving the original value of the number!

Decimal Places

Any number to the right of the decimal point is your decimal figure. Since these numbers can be recurring, we may round-off our figures.

Rounding-off Decimals

Rounding off decimal figures is no different from rounding your numbers to the nearest tens or whole numbers. If a math problem requires you to round-off a figure to a specific decimal place (d.p), we round-off base on the number right next to that decimal place value.

Consider 2.815 to 2 d.p:

2.81⑤6 = 2.82

> Our 1st decimal figure is 8, which means that 1 is the second decimal place value. We check the number to its right, if it is 5 & up, we add 1 to our 2nd decimal place value. If the number is 0 - 4 we ignore all decimal values beyond the 2nd decimal place value.

Significant Figures

Unlike decimal places, a number before or after the decimal point maybe considered to be significant. There are a few things you need to know when determining if a number is significant or not.

- All leading zeroes are insignificant (zeroes to the left of nonzero digits)
- All Non-zero digits are significant
- Zeroes in between two non-zero digits are significant
- Trailing zeroes are only significant if the number is a decimal or has a decimal point.

If any of the last (3) above criteria is met, then that digit in that number is significant!

00<u>129</u> = 3 s.f   0.00<u>352</u> = 3 s.f   <u>20009</u> = 5 s.f

<u>4</u>000 = 1 s.f   <u>4000.</u> = 4 s.f   0.00<u>50</u> = 2 s.f

When given a figure to be written in significant figures (s.f), we identify our significant figures first. We round off based on the value after that significant figure.

<u>Example:</u>

Write to 3 s.f:

$$\overset{1\ 2\ 3\ \ 4}{0.014\ \underline{5}\ ⑥} = 0.0146$$

> Using the rules of significant figures, we see that we have 4 significant figures. Since we are rounding off to 3 s.f, we check the number next to the 3rd s.f. If it is 0-4, we ignore all the digits to the right of the 3rd s.f. However, if it is 5-9, we add 1 to the 3rd significant figure and ignore all the digits to its right.

Practice Activity

a). Calculate the exact value of:

$$\frac{2.4 + 1.76}{4.4 - 3.1}$$

b). Calculate 9.72 x 12.05; write your answer in Standard form

c). Evaluate $(0.35)^2$ - 0.03 x 0.8 to 2 d.p.

d). Write 4768 correct to 3 s.f.

e). Express 0.8909 in:
   i. in standard form
   ii. correct to 2 s.f.

f). Determine the exact value of.

$$\sqrt{\frac{13.5}{0.33}}$$

   ii. to 3 d.p

g). Calculate the value of
   i. $(4.2 \times 10 ^\wedge 4) \times (5 \times 10 ^\wedge -3)$
   ii. Express your answer in standard form

h). Evaluate $2.36_2 + 4.1_2$ to 2d.p

i). Write exactly the value of 0.428 x 2.75 to:
   i. 2 significant figures
   ii. in standard form

j). Express 0.0402 ÷ 0.71 to:
   i. correct to 2 d.p
   ii. correct to 3 s.f
   iii. in standard form

PERCENTAGES

With the foundation being well laid in the previous lesson, you should be able to grasp this one perfectly. Let us look at our definition of the word 'Percentage':

Per => For Each
Cent => 100 (as in century, centimetre and cents)

Sometimes 'Per' is represented by "/" as a symbol. Mathematically speaking the term would denote that we should divide by "cent"; which is 100. Let's understand our definition. When we say 4%, what we are actually saying is " 4 for each 100 or 4/100 ". Now there is only just one 100, and 4 into 100 = 25; so that's 1/25.

The main thing students are to get is anything in percentage (x%) is expressed as $\frac{x}{100}$. We may want to change a number into percentage. When trying to convert a number to percentage we do ' number x 100 ', or $\frac{x}{y}$ x 100. So 1/2 % or 1/2 for each 100 means, the 1/2 of 100; which equals 50%. Understanding all this, it should make solving our math questions clear enough to solve.

Converting Numbers to Percentages:

1. 2 converted in percentage is 2 x 100 = 200%
2. 0.35 converted in percentage is 0.35 x 100 = 35%
3. 3/4 converted in percentage is 3/4 x 100 = 75%

Just simply times the number by 100 and it is converted.

Converting Percentages to Numbers:

1. 200% converted as $\frac{200}{100}$ which equals 2
2. 35% converted as $\frac{35}{100}$ which equals 0.35
3. 75% converted as $\frac{75}{100}$ which equals 3/4 or 0.75.

Simply divide the number by 100 and it is converted.

Notice that in our conversions, it should be highlighted that making a number written in percent form doesn't change the original value of the number. When converting numbers to any unit (percentages or metric units) you are neither adding or subtracting to the number.

See example below:

5m is converted as 5000 in mm, 500 in cm and 50 in dm. Although the number itself may differ in a numerical sense, the metric units get stronger. Ideally, the object at 5m measures the same length every time without losing or adding length if measured in these units.

Increasing a Number by percentage:

We can increase a number by a set percentage in two ways

- Primary level:
    - find the percentage of the number.
    - Then add the result to the number.

Increase 125 by 20%

$$\frac{\overset{1}{\cancel{20}}}{\underset{1}{\cancel{100}}} \times \frac{\overset{25}{\cancel{125}}}{1} = 25$$

The total amount is 125 + 25 = 150

- Higher Level:
    - Add the percentage that the number is being increased by to 100%
    - Convert the percentage to a real number
    - Multiply the number to be increased to your result

Increase 125 by 20%

20% + 100% (Represents 125, as 125 is 100% of itself)

120% is expressed as $\frac{120}{100} = 1.2$

125 x 1.2 = 150

We could have expressed our solutions by having 20% as $\frac{20}{100} = 0.2$ and just add 1 = 1.2. In short, 100% is equivalent to 1. This higher level calculation may seem more than the old primary way. However, after understanding how it works, you will realize how faster and easier math becomes.

How about reducing a number by a percentage?

Reducing a number by a percentage:

- We could do it at the primary level and find the percentage of the number and then subtract it from the number to be reduced. This is similar to what we did to increase a number by a percentage, the only difference is that we would subtract our results.

Decrease 150 by 25%

$$\frac{\overset{1}{\cancel{25}}}{\underset{\underset{1}{\cancel{4}}}{\cancel{100}}} \times \frac{\overset{37.5}{\cancel{150}}}{1} = 37.5$$

The difference of 150 - 25% is equal to 150 - 37.5 = 112.5

- Higher Level:

Decrease 150 by 25%

$100\% - 25\% = 75\% = \frac{75}{100} = 0.75$

Or we could have 25% as $\frac{25}{100} = 0.25$. Then we would say 1 - 0.25 = 0.75

Our final results would be 150 x 0.75 = 112.5

If the price of an item Increases by inflation, or decreases on discount/sale by a set percentage, we can apply these methods respectively to find the final sale price of the item.

What if you have been given a problem to obtain the original price on an item sold at a value with the tax rate already added to it?

Example:
A price of $480 is the final sale of an item after a 5% tax, what is the cost of the item?

To obtain the original cost amount after a given percentage is added or subtracted from a number, we follow the following formula:

- **Increase Original Cost**

$$\frac{\text{Final Price x 100}}{(100 + \text{Increase \%})}$$

- **Decrease Original Cost**

$$\frac{\text{Final Price x 100}}{(100 - \text{Decrease \%})}$$

Let's evaluate $525 inclusive of 5%.

Find the exact cost of the item:

Based on our statement, we can see that 5% was added to the price of an item. Therefore this question require the increase formula for our solution.

final price = $525
Percentage added to item cost = 5%

So our formula is expressed as:

$$\frac{525 \text{ x } 100}{100 + 5} = \frac{52,500}{105} = \$500$$

When we add the percentage and apply our formula, notice how our answer is less than the final cost of the item. This is because we are finding the exact cost of the item before the item tax was added.

Let's find our cost price after a reduction of a certain percentage.

An item costs 'x' amount. A customer comes to the store and decided to purchase the item. He thinks the item is a bit too pricey and asks for a discount. The store decides to sell him the item with 20% discount. The man pays $300 for the item. How much was the cost of the item?

Using 'x' as our original cost:

$$X = \frac{300 \times 100}{100 - 20} = \frac{30,000}{80} = \$375$$

Notice in this example how our result is greater in value than the final cost of the item. This is because the cost of the item was reduced from what it was initially to a final sale price.
These formulae are not limited to money but in any math problems that requires you to find the original values.

Example:
The International Stadium this year had 6,000 patrons seated, which is 25% more in comparison to last year when it hosted the football games. How many patrons were seated last year?
(The correct approach to this question is to use the original cost increased formula)

$$\frac{6000 \times 100}{100 + 25} = \frac{600,000}{125} = 4,800 \text{ patrons}$$

Many students would have approached this question by finding 25% of 6,000 and subtract it. This would make 4,500 be the answer, but try adding 25% to 4,500 and see if you get 6,000 to be 25% increase of 4,500.

Buying and selling

Consumer Arithmetics simply has to do with the logics or calculations relating to buying and selling. The consumer is the person buying or utilizing a product or service. Arithmetics is the mathematical side of getting the product or service and how to utilize it effectively.

Terms and formulas to know:

Cost-Price
The original price of the produce/service.

Selling-Price or Marked-Price
is the final price set on the produce or service to be purchased. Also called the retail price.

Profit 1
If the selling price is greater than the cost price then you have made a profit.
(Formula is therefore SP - CP).

Profit 2
If given the percentage profit made and the cost price, you can obtain the actual profit made.

$$\text{Profit} = \frac{\%\text{Profit}}{100} \times \text{Cost-Price}$$

Loss
If the selling price is less than the cost price then a loss is incurred.
(Formula is CP - SP).

Percentage Profit/Loss
This is expressed as the profit made of the cost price times 100.

$$\text{Formula} = \frac{\text{Profit/Loss}}{\text{Cost-Price}} \times \frac{100}{1}$$

Example:

- Calculate the profit or Loss as a percentage on an item marked at $250 after ordering it for $200.

Profit = Selling Price - Cost Price

$$\frac{250 - 200}{200} \quad \text{x} \quad \frac{\overset{}{\cancel{100}}}{1} = \frac{\overset{25}{\cancel{50}}}{\underset{1}{\cancel{2}}} = 25\%$$

In solving any worded question, the right thing to do is to make note of your data. The first thing here is to make note that our selling-price is bigger; so therefore a profit was made. We then find the profit that was made and identify the cost-price of the item. We plug in the values from our data into the formula and then evaluate our answer.

This formula may also be used or substituted in given scenarios to find the percentage increase or decrease of a given thing or situation.

Example:

In the 1900s, the stadium was filled on an average of 20,000 patrons. In 2010, after a massive pandemic attack, the average patrons fell to about 16,000. Calculate the difference in percentage.

Difference or Loss = 20,000 - 16,000 = 4,000

(For this question, the loss formula will be substituted to provide the answer)

$$\frac{\overset{1}{\cancel{4000}}}{\underset{\underset{1}{\cancel{5}}}{\cancel{20000}}} \quad \text{x} \quad \frac{\overset{20}{\cancel{100}}}{1} = 20\%$$

Formula =	Diff	x	100
	Prev Amt		1

Practice Activity

a). An article is valued at $2,499, for how much money should a merchant sells this item for to make 20% profit?

b). A local store gives 15% discount on all items. How much would a customer pay for an item valued at $750.

c). If 20% of a sum of money is $360, what is the sum of money?

d). If 24 ounces of a cake represents 25% of its mass, what is the mass of the entire cake?

e). John has $42.00 in his pocket, he is planning to spend $20.00 for his lunch and $14.00 for his traveling. What percentage of his money is he planning to use for his traveling?

f). In a bucket containing 360 plums 216 are ripe, calculate:

  i). What percentage of the plums are ripe?

  ii). If 75% of the apples were ripe, how many apples would that represent?

g). In a sale, the prices were reduced by 12%. What is the original cost of an item that is now being sold for $2,200

h). A man sold an article for $50 dollars thereby making a profit of 25%. How much did he pay for the article?

SIMPLE AND COMPOUND INTEREST

When you borrow a loan from the bank, an interest rate is applied. The longer you take to repay the bank is the more debt incurred. This is because the value of interest grows over time. You may also make deposits or investments at a set interest rate to reach a target savings amount by a set time in the future. We calculate the value of Interest by I = PRT, where

I = Simple Interest

P = Principal (The amount Invested/borrowed)

R = Interest rate in percentage, usually a yearly rate.

T = The time allotted for the investment or loan.

Since the rates is a percentage we may express this formula as follows I = PRT/100. This formula may also be transposed to solve different individual parts.

$$I = \frac{PRT}{100} \qquad P = \frac{I \times 100}{T \times R} \qquad R = \frac{I \times 100}{P \times T} \qquad T = \frac{I \times 100}{P \times R}$$

Consider the following:

- What is the interest on an investment of $2,500 at 8% per annum for 2 years?

Since the problem explicitly asks for the interest, I therefore use the simple interest formula.

$$I = \frac{2500 \times 8 \times 2}{100} = \$156.25$$

- What is the interest when invest $500,000 at 12 1/2% per annum in 1 3/4 years?

We can start by converting our Rate and Time from mixed fractions to improper fractions.

$$\text{Rate} = \frac{25}{2} \quad \text{Time} = \frac{7}{4}$$

$$I = \frac{\$500,000 \times 25 \times 7}{100 \times 2 \times 4} = \$109,375$$

When dealing with simple interest questions, students should be mindful of what the question is asking to solve. Sometimes questions may ask you the actual amount of money saved after the period. Other questions maybe that we need to solve the other 3 Values (P, R or T).

Let us look at some example questions:

1). What is the amount invested at 5% per annum in 3 years to give an interest of $525?

$$P = \frac{525 \times 100}{3 \times 5} = \frac{52,500}{15} = \$3500$$

If we do the simple interest formula, and use our principal of $3,500 for the investment for 3 years at 5% per annum, you should get $525 interest.

2). How long should $5,000 be invested for inorder to accumulate $450 interest at 2 1/4% per annum?

$$T = \frac{450 \times 100}{5000 \times 2.25} = 4 \text{ Years}$$

Since our rate is per annum, it means that all my data is given for a yearly period.

3). What is the rate percent on an investment of $2,700 for 3 1/5 years, if the interest amounts to $840.

$$R = \frac{840 \times 100 \times 5}{2500 \times 16} = 16.5\%$$

Notice how we converted our time from Mixed fractions to Improper fractions (16/5), and then reciprocate it to find our solution. We could have just written it out in decimals, but I want this lesson to be flexible in solving mathematical problems.

Compound Interest

You may be asking yourself the following questions at the moment:

- What is compound Interest?
- How does it differ from Simple Interest?

<table>
<tr><td align="center">**Simple**</td><td align="center">**Compound**</td></tr>
</table>

What is the Interest on a $20,000 investment for 2 years at 12% per annum?	What is the Interest on a $20,000 investment for 2 years at 12% per annum compounded annually?

Simple side:

$$I = P R T$$

$$I = \frac{20000 \times 12 \times 2}{100} = \frac{480000}{100} \quad \text{①}$$

$$= \$4,800$$

At the end of the period the savings would amount to $24,800 ($20,000 + $4,800).

Compound side:

$$I = P R T$$

$$I = \frac{20000 \times 12 \times 1}{100} = \$2,400 \quad \text{①}$$

$$20,000 + 2,400 = \$22,400.$$

$$I = \frac{22400 \times 12 \times 1}{100} = \$2,688 \quad \text{②}$$

$$22400 + 2688 = \$25,088.$$

At the end of the period the savings would amount to $25,088.

Have you noticed any differences here? If yes, what are those differences? It should be clear that we have two different amount by the end of the two years in our savings account. With Compound Interest we were able to save $288 more than our Simple Interest. We also should note that we calculated our interest twice within 2 years for the Compound Interest and only once for the Simple Interest.

When interest is compounded, it means that the interest has been added to the principal at a set amount of times within a year. An interest may be compounded daily (365 times per year), monthly (12 times per year), annually (1 time per year) or even Quarterly (4 times per year) within a given time period.

If we were to calculate the compound interest every time using the way it was just explained, it would be very tedious for us. For instance, if the problem becomes complex like a daily compound interest for even 1 year, we literally would have to do the calculations 365 times to see our results. There is another formula to do both interest rate with the answer being the total amount of money accumulated over the period 'A'.

We represent the amount of time the interest is calculated with a year by 'n' in the formula:

$$A = P \left(1 + \frac{r}{n}\right)^{nt}$$

- A = Final amount accumulated after a set period or simply, the accrued sum of money.
- P = Principal amount of money where interest is being accumulated on.
- R = Rate percent increase on the principal within the period
- T = Time period or years allotted for the investment.

In our previous working out using simple interest formula, our 'T' takes the meaning of 'n'. We were given an annual compound interest. That means at the end of each year, the total amount is calculated for the beginning of the following year; so our principal becomes the accrued sum of money for the new year.

Let us see how our formula works:
A = ?   P = $20,000   R = 12% or 0.12   n = 1 (Annual C.I.)   T = 2 years

$$A = 20,000 \left(1 + \frac{0.12}{1}\right)^{1 \times 2} = 20,000 (1.12)^2 = \$25,088.$$

(To calculate our interest, we do A - P = I. So that's $25,088 - $20,000 = $5,088).

*You might feel the need to use this formula for your Simple Interest: $A = P \left(1 + \frac{rt}{100}\right)$.

Did you just noticed the 1 being added to 12%? Does it look familiar? It is like when we increase a number by 12% we add it to 100%, resulting 112%. If we were to convert our percentages to a number we get 0.12 + 1 which gives us 1.12. This is because interest is calculated by the rate; We are in creasing our principal by a rate of 12% each year for 2 years!

Depreciation:

This is analogous to compound interest, the only difference is that the result represents a decrease. We use depreciation when something loses value over a set period of time, like machinery, vehicles and so on.
We may use the same formula as our interest as D = PRT, where D, represents the Depreciated value on an item.

Let us look at an example:

What is the value of a vehicle that experienced a 8% depreciation each year over a period of 3years if its initial value was 900,000?

$$D = \frac{900000 \times 8 \times 1}{100} = 72,000$$
($900,000 - $72,000 = $828,000)

$$D = \frac{828,000 \times 8 \times 1}{100} = \$66,240$$
($828,000 - $66,240 = $761,760)

$$D = \frac{761,760 \times 8 \times 1}{100} = \$60,940.80$$
($761,760 - $60,940.80 = $700,819.20)

This is very similar to what we did when solving our problem for Compound Interest using the Simple Interest Formula.

Let us do the same math using this short hand formula below:

$V = P \left(1 - \dfrac{r}{100} \right)^3$, Where V is the value of the commodity after the given period.

We now say:

$V = 900{,}000 \, (1 - 0.08)^3$
$= 900{,}000 \, (0.92)^3$
$= 900{,}000 \times (0.778688)$
$V = \$700{,}819.20$

Practice Activity

a). A father decided to open up a savings account for his child. He decided to invest $24,000 at the birth of his daughter, the account pays 8% simple interest.

    i. How much interest will be made on the account after 10 years?
    ii. What is the value of the account when his daughter turns 18?

b). Calculate the total sum of money in a savings account after 2 years, if an investment of $50,000 was to be deposited at 14.6% per annum compounded daily.

c). If man invests $132,000 for 1 year at a rate of 6% per annum compounded quarterly, how much money would he save?

d). Calculate the value on a motorbike that depreciates at 10% each year over a period of 8 years, if its initial value was $350,000.

e). A washing machine is bought for $12,500 and depreciates in value at 8% per year. Calculate its value after 5 years.

Ratio

When we divide quantities into proportions by a constant variable, we record it in the form of ratios. Ratio is a mathematical data that is used to show the relationship between 2 or more quantities. In this lesson, you will find that ratios are another way of expressing proportions in the form of fractions of a related quantity. Ratios are expressed as x:y:z; where x + y + z equals the total quantity, i.e. all of these values (x, y and z) are a fraction of the total quantity.

Let's consider the ratio 4:8:6. When we add each part we get 18 units of the divided quantity. We can express this as follows:

$$\frac{4}{18} + \frac{8}{18} + \frac{6}{18} = \frac{18}{18}$$

Each fraction of the fractions above can be broken down to lower terms if we cancel by 2.

$$\frac{2}{9} + \frac{4}{9} + \frac{3}{9} = \frac{9}{9}$$

Although 3/9 could be broken down to its equivalent (1/3), we must have a consistency in our denominators to write down our values as x : y : z. So our ratio could have been expressed as (2 : 4 : 3; a sum of 9 units). Our ratio remains consistent as it is still equivalent to the previous 4:8:6. Try multiplying each proportion of the ratio 2 : 4 : 3 by 2 and see.

As discussed ratios are forms of fractions, but same is true in reverse.
The ratio 40 : 80 can be written as:

- $40 : 80 => \dfrac{\overset{1}{\cancel{40}}}{\underset{2}{\cancel{80}}} = \dfrac{1}{2}$

- $\dfrac{1}{2} \overset{2}{} => 1 : 2$

Let us examine a few ratio type questions.

On maps, ratios are given to measure the actual distance as 1: 15,000. If we were to calculate the actual distance, in km, between two points in the city which lies 87 m on the map, we approach this question as follows:

Since our ratio is 1 : 15, 000, which means that for every unit 'm' on the map, we multiply 87 by 15,000 to equal the actual distance. In essence, what we are really saying is that 1 unit (1metre) values 15, 000 times in the actual distance or 1unit = 15, 000 units (Actual distance). Since locations on the map are metres a part, the actual distance is 15,000 times more. Therefore 87 units on the map measures 1,305,000 in metres (actual distance), and converting m in km, it is 1,305 km, actual distance.

Sharing according to a given ratio

- Let us share $2,000 in the ratio 5:9:6. (5 + 9 + 6 = 20).

We want to know 5 units of 2000, 9 units of 2000 and 6 units of 2000 dollar. Now remember that each units of our ratio is a fraction of the whole 20units. That means if we want to find each units of 2,000 we may express our equations as follows:

$$\frac{5}{20} \times \frac{\overset{100}{2,000}}{1} = \$500 \qquad\qquad \frac{\overset{100}{2,000}}{\underset{1}{20}} = \$100$$

$$\frac{9}{20} \times \frac{\overset{100}{2,000}}{1} = \$900 \qquad \text{OR} \qquad \text{(Therefore 1 ratio unit = \$100)}$$

So we say:

$$5 \times 100 = \$500$$

$$\frac{6}{20} \times \frac{\overset{100}{2,000}}{1} = \$600 \qquad\qquad 9 \times 100 = \$900$$

$$6 \times 100 = \$600$$

$500 + $900 + $600 = $2,000

The truth is that the whole sum of money to be shared, was divided into fractions. Each figure representing the amount shared in ratio (500 : 900 : 600). We can then break down this ratio into smaller measurable quantities by dividing with the HCF (100); 500/100, 900/100 and 600/100, which is equivalent to 5, 9 and 6 respectively, which in ratio is 5 : 9 : 6.

This makes it clear that our sum of the ratio (20) is also an equivalent fraction of 2,000. This provides us with a better understanding as to how ratios work. We can then apply our equation more smoothly by the following steps: 2,000/20 (total to be shared divided by the sum of each shares in the ratio).

We can therefore say that 1 ratio unit is equivalent to 100 units of the total shares. So we have 1 : 100 or 1 ratio = 100 shares. Now a share of 5 ratio units can be expressed as:

5 x 100 = 500   or   5 ratio = 500 shares
and 9 x 100 = 900 shares
and also 6 x 100 = 600 shares

We simply just multiply our ratio by the value of 1 unit, to find each shares. This is exactly what was done in our second example.

Alternative work with Ratios

What if you are not given the total to be shared, but only a part of the ratio? How would you find the other part?

Consider the following example:
- A piece of string is divided in the ratio 2 : 6. The larger is 48 cm long find the length of the other piece.

Most students see this question and start by adding the ratio (2 : 6) and divide it into 48. They would get 6 for their answer. However, this is the wrong approach to this question! To solve any ratio question where we need to find the value of the shares of the ratio, we must first find the value of 1 unit ratio. The difference with this question from the previous is that we are not given the total that was shared, but instead, if we read closely we are given the value of one of the ratios. The larger share in the ratio is 6 and the problem stated that the ratio of 6 values 48 units(cm). Therefore we evaluate that value of 1 ratio as follows:

$$\frac{8}{\cancel{48}} = 8 \text{ units (cm)}$$
$$\overline{\cancel{6}}$$
$$1$$

> Therefore 1 unit of our ratio is equivalent to 8 cm of the actual length of the whole string.

We can express this information as a ratio 6 : 48 units is equivalent to a ratio of 1 : 8 units(cm).

We solve for the other part by doing:
The ratio 2 x 8cm = 16cm. The total ratio can be calculated by adding the two parts 16cm + 48cm = 64cm.

We could have worked out the total length of the string by first finding the sum of the ratio, 2 + 6 = 8. Then we do the sum of the ratio times 8cm; 8 x 8cm = 64cm.

Another way we could have done this is as follows:
We could have 6 : 48 being expressed as 6x = 48; where x, represents the value of '1unit'. We would do the same exact thing as we did previously, having x = 8cm. We then have 2x : 6x, which would equal 2 (8) : 6 (8). So our final answer would be 16cm : 48cm, which equals a total of 64cm.

Finding the ratio of shares

If you were to go over the previous activities, you should not have a problem solving the following:

- A sum of $2,500 was shared between John, Anna and Nancy, such that John received $800, Anna received half and Nancy received the remainder. Find the ratio in which the money was shared.

The shares would be divided in the ratio J:A:N, where J = $800, A = $1250 and N = 450; so 800 : 1250 : 450.

To find the ratio in which the sum of money was divided we have our solution written as:

$$\frac{800}{2500} = \frac{8}{25} \times \frac{2}{2} = \frac{16}{50}$$

$$\frac{1 \times 25}{2 \times 25} = \frac{25}{50}$$

$$\frac{450}{2500} = \frac{9}{50}$$

To find the ratio, all the denominators must be the same as discussed earlier. We achieve this by finding the LCM between the three denominators 25, 2 and 50; and convert the fractions to their equivalent accordingly.

The sum of $2,500 was therefore shared in the ratio 16 : 25 : 9.

Let us try to do another ratio type problem:

- A sum of money is divided amongst 3 girls Anna, Barbara and Christine in the ratio 5:3:2. If Barbara received $400 less than Anna, calculate the sum of money that each girls would receive.

In our previous exercises, we covered how we find 1 unit ratio, then multiply each proportion by the value of 1 unit. So we say 5x : 3x : 2x; where x, is the variable that represents the value of 1 unit ratio.

Now, unlike previous exercises, we are not given the values of our ratio. However, instead we are told that the difference between Barbara and Anna amounts to $400. We know that Barbara receives 5x and Anna receives 3x, so our equation should look like this:

5x - 3x = $400.

Therefore we should have 2x = $400. This is further evaluated as:

$$x = \frac{400}{2} = \$200$$

Therefore a ratio of 1 is equivalent to $200. Now we are able to calculate how much money was shared and how much money does each individual received.

Anna received 5 (200) = $1000
Barbara received 3 (200) = $600
Christie received 2 (200) = $400

We see where $1000 - $600 = $400 confirming our answer. The total amount of money is caculated by adding the sum of the ratio times 200 (10 x 200), which gives us $2,000; also you may just simply add the individual shares: 1,000 + 600 + 400 = $2,000.

Practice Activity

a). An entire cake was divided in (3) three parts in the ratio 1:5:4, the smallest piece was 4 ounces. What was the total weight of the cake?

b). Water was placed into (2) two containers in the ratio 6:11. The one with the greater volume contained 44 litres. How much water was in the two containers combined?

c). A metal is made from copper, zinc and lead in the ratio 13 : 6 : 1. The mass of zinc is 90 kg. Calculate the mass of the metal.

d) Share $3,500 among Matthew, Kiara and Shara so that each of the two girls receives thrice as much as Matthew.
   i. Find the individual shares
   ii. What is the ratio that the money was shared?

e). A sum of money is divided amongst (3) three persons Nieka, Lewy and Simon in the ratio 1:3:5. If Nieka received $500 less than Lewy, calculate the sum of money that each individual would receive.

DIRECT & INVERSE VARIATIONS

What are direct and inverse variations? A lot of topics are done with the concept of direct and inverse variations, and without even knowing what they mean we are sometimes able to apply the correct formula to solve our problems. In this lesson we will now understand the fundamentals of some math problems and how to solve them more accurately.

Direct Variation

If a variable 'y' increases as another variable 'x' increases, and also decreases as 'x' decreases, we say 'y' varies directly as x. We may also say that 'y' is directly proportional to 'x':

On a graph, this information would look like this:

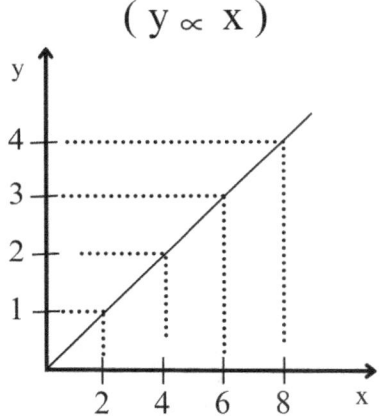

$(y \propto x)$

By looking at our graph, we see that when y increases by 1, x increases by 2. The relationship between y and x is a constant variable, k, which equals 1/2 or 0.5. Note that k also represents the gradient of the straight line.

The formula for calculating direct inverse is given as: y = kx, where 'k' is a constant of proportionality. In order to evaluate our equation, we need to determine value of our constant 'k'.

Let us examine the theory:

If ten 6 boxes of Orange Juice cost $60, how much would 24 boxes cost? How many boxes can be bought with $1,500?

y = boxes and x = cost (the variables y and x could be either way).

We want to obtain the value for our constant variable, k, first:
Formula (y = kx)

$$6 = k \ \$60$$

$$\frac{\overset{1}{\cancel{6}}}{\underset{10}{\cancel{60}}} = k = 0.1$$

Part 1 solution:

Now that we have our constant variable, k, we can evaluate the value of 24 boxes of Orange Juice. We simply plug in the values of our variables:
Formula (y = kx)

$$24 = 0.1X$$

$$\frac{24}{0.1} = X = 240$$

Therefore the cost of 24 boxes of Orange juice is $240.

Part 2 solution:

We can also evaluate how many boxes can be bought with our x variable, $1,500.
Formula (y = 0.1x)

$$y = 0.1 \times 1,500$$
$$y = 150$$

Therefore $1,500 is able to purchase 150 boxes of Orange Juice.

Let us look at another example:

Y varies directly as x^2. If 'y' is equal to 48 when x is equal to 4, find the value of x where 'y' is equal to 75 and the value of y when x is equal to 3.

The above proportionality is written as ($y \propto x^2$). The formula is therefore given as $y = kx^2$.

Not knowing the formula for direct variations, students would normal resort to solving this by doing:

Part 1 solution:

$$\cancel{48} = \cancel{4^2} \; \Longrightarrow \; 48x^2 = (4^2 \text{x } 75)$$
$$\cancel{75} = \cancel{x^2}$$

$$48x^2 = 1200$$
$$x^2 = \frac{1200}{48} = 25$$
$$x = \sqrt{25} = 5$$

Although this works very well, when we get to inverse variations, this way of evaluating your answers will not work. With understanding of direct variations we can solve part 2 of the problem.

Part 2 solution

Using our formula ($y = kx^2$)

$$48 = k4^2$$

$$\frac{\cancel{48}^{\,3}}{\cancel{16}_{\,1}} = k = 3$$

We now solve part 2 of the problem by inserting our constant, k, and our x values. We insert the values according to our given formula $y = kx^2$.

$$y = 3(3)^2$$
$$y = 3\,(9)$$
$$y = 27$$

(Feel free to solve part 1 of the problem by using the constant variable 3 in the formula).

Inverse Variation

We say that 'y' varies inversely as 'x' if when 'x' increases, 'y' decreases and also when 'x' decreases, 'y' increases. We say that 'y' is inversely proportional to 'x', $y \propto \frac{1}{x}$. Therefore our formula is $y = \frac{k}{x}$.

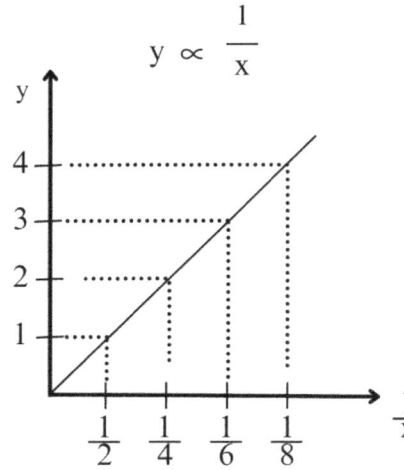

Looking at the data from the graph, we see where as y increases x decreases. The graph looks similarly to direct variations, but with each value written in their inverses.

Now with this understanding let us try solving a problem:

If it takes the strength of 30 men to complete a task in 18 days, how long would it take 50 men? Also how many men would be required to complete the same task in 7 days?
(Using $y = \frac{k}{x}$), where y = No. Men and x = No. Days.

$$30 = \frac{k}{18}$$

$$30 \times 18 = k = 540$$

Part 1 solution:

The strength of 50 men will complete the same task in:

$$50 = \frac{540}{x}$$

$$50x = 540$$

$$x = \frac{\overset{10.8}{\cancel{540}}}{\underset{1}{\cancel{50}}} = 10.8 \ (11 \ days)$$

We didn't necessarily round-off our answers to the nearest whole number. It is only logical to say that the work went over and was accomplished in the following day.

Part 2 solution:

The same task can be done in 7 days with:

k = 540;

$$y = \frac{540}{7} = 77.1 \text{ or } 78 \text{ men}$$

Notice here that we round our answer to the nearest whole number even though we have 1 tenths. If we used 77 men, the task would be accomplished in the 8th day just like our previous example.

Let us do one more example:

'Y' varies inversely as x^2. If when y equals 12 the variable x equals 4, find the value of y when x equals 1.

Use $y = \frac{k}{x^2}$;

$$12 = \frac{k}{4^2}$$

$$12 \times 16 = k = 192$$

$$y = \frac{192}{4} = 48$$

It may be easy to forget that your variable x is squared. It is important to always look back at the formula when solving these equations. If you were to plot your graph, using your answers, you will see that as y increases, x decreases and vice versa.

Solution:

$$y = \frac{48}{1^2} = 48$$

Practice Activity

a). If m varies directly as v^2, and m = 2 and v = 3. Calculate the value of m when v = 6.

b). Given that q varies directly as p, using the values of p and q in the table, calculate the values of a and b.

p	2	8	a
q	6.1	b	1.2

c). Given that y varies inversely as x^2, and that y = 3 when x = 2, calculate the value of y when x = 3.

d). Given that y varies inversely as x, and y = 8 when x = 2, calculate the value of y when x = 32.

e). If S varies directly as (r + 1), and S = 8 when r = 3, calculate the value of r when S = 20.

f). Given that p varies directly as r^3, and p = 4 when r = 2, calculate:

   i. the value of p when r = 0.2
   ii. the value r when p = 62.5

g). y varies inversely as $(x^2 - 1)$. When x = 3, y = 15. Find the value of:

   i. x when y = 5
   ii. y when x = 2

FOREIGN EXCHANGE

 Foreign Exchange has to do with the conversion of one currency to another by a rate of exchange. A rate of exchange is given in the form of a ratio, where each currency pairs show the value of one currency at $1.00 expressed in another. For example: US $1 : EC $2.70 which is interpreted as $1.00 US is equivalent to $2.70 EC.

 These rates of exchange is usually seen in Banks, Western-Unions, Money-Grams, Cambios and other business places, where trading or foreign exchange is done. You will see different pricing for buying and selling, or bid and ask pricing. The bid or buying price is usually less than the ask price. It is the price in which a company is willing to buy a currency from you in exchange of a local or foreign currency. The ask or selling price tells you how much the company is willing to trade or sell you $1 foreign currency for in exchange of the local or foreign currency. The difference between the ask price and the bid price is referred to as the 'spread', which is the a profit made from exchanging currencies.

 Using our rate of exchange US $1.00 : EC $2.70, we say that $2.70 EC is required to buy US$1.00. With this understanding you should be able to exchange any sum of money in either currency.

Let us do an example:
Convert 2,700 ECD to USD

Since there is 1 USD in every 2.70 ECD, it means that we need to find out how many 2.70 ECD is in 2,700 ECD to give us our USD value.

 2700 ECD ÷ 2.70 ECD = 1000 USD

Since there is 1,000 of $2.70 ECD (equivalent to $1 USD) in $2,700 EC, $1,000 US is equivalent to EC $2,700. Take note also how the ECD cancelled each other, giving a new currency.

Let us try to convert
$500 US to EC$:

Rate of Exchange
 $1 US = $2.70 EC
 500 x 1 = 500 x 2.70

$500 US = $1,350 EC

Since $1 US is equivalent to $2.70 in EC, we multiply the rate by the amount of USD. We can see also where the half of USD $1,000 (equivalent to 2,700 ECD) is $500, which is half of its equivalent (ECD 2700/2 = ECD 1350). Also notice that for one operation we divide and for the other we multiply.The question therefore lies when do we multiply from when to divide?

 If you were to take a minute and look at what is happening, it could come clear to your minds to notice something. Any amount of money we have in any currency pair, the other currency moves along. So if we have $500 USD we have $ 1,350 ECD, and if we have 1000 USD the ECD is 2,700. Does this seem familiar to you? So the value of money in one currency increases as the value of money of the other currency increases, and if the value of money decreases, so does the value of money decreases of the counterpart currency! This is therefore an example of 'Direct Variation'.

Let us test the theory:
$y = kx$, where y is USD and x is ECD.

2700 ECD to USD

Exchange rate:
$1.00 US : $2.70 EC
1.00 = k x 2.70
$\frac{1}{2.7}$ = k = 0.37037037

Part 1 solution:
USD = k x ECD
USD = 0.37037037 x 2700
Answer = $1,000 US

Part 2 solution:
Convert $500 US to ECD:
y = USD and x = ECD

y = 0.37037037x
500 = 0.37037037x

$$\frac{500}{0.37037037} = x = \$1,350 \text{ EC}$$

You see?! we were able to use the principle of direct variation to obtain our exchanges! This removes the questioning of "should I multiply or should I divide" from our minds. A key practice that we should do is always represent as much decimal values as possible when converting currencies as you have seen in my examples. Currencies or money in general are calculated more accurately when there are more than enough decimal values.

Let us look at another problem:
Given the rate of exchange

US	$1 : EC $ 2.70
TT	$1 : EC $ 0.40

Convert TT$648 in USD.

When we look at our exchange rate, we don't see a pair for $TT to USD. However we see where we may be able to convert $648 TT to ECD and use that result to convert to USD using the ECD to USD currency pair.

Step 1 solution:
Converting TT $ 648 in EC $ using the rate of exchange TT$1.00 is equivalent to EC $0.40.

TT$ 648 x 0.4 ECD = $259.20 EC

This is where students get stuck, as they become confuse on what to do next. While it may be clear to some that something should be done with the $259.20 EC and the USD to ECD pair, the real confusion is the indecisiveness as to whether they should multiply or divide to solve the equation.

<u>Let us use our direct variation formula to remove all confusions:</u>

y = 0.37037037x

y = USD and x = ECD

y = 0.37037037 x 259.20;

y = $96

Therefore TT$ 648 is equivalent to $96 US. This totally removes the questioning from your minds, as to whether or not you should multiply or divide!

Revaluation and Devaluation

Revaluation

When a currency's demand increases, it tends to gain power and stability. This increases its value on the Forex market. The currency therefore is revaluated against its counterparts. This means that it takes more of the other currency to purchase the revaluated currency.

<u>For example:</u>

Given the rate of exchange 1 CAD is equivalent to 110 JMD, if CAD revaluated at 20%, then $1 CAD = $132 JMD. It now takes more JMD for the price of 1 CAD.

<u>We revaluate a currency as follows:</u>

Revaluation is an increase of 20%, therefore CAD has increased in power by 20% against its counterpart. It must be clear in our minds that if CAD has revaluated by 20% against its counterpart, it means that it takes 20% more of what JMD originally was to purchase the CAD currency.

20% + 100% = 120% = 1.12

CAD $1.00 : 110 JMD x 1.2 = CAD$ 1.00 : 132 JMD. (We can see from this solution that it takes 20% more of what JMD used to take inorder to buy $1 CAD, hence the CAD revaluated). We could have just simply find 20% of 110 and then add the results when we are revaluating a currency, but that doesn't work for devaluation.

Devaluation

In contrast to Revaluation, if a currency takes less of what it used to worth to buy its counterpart currency, then we say that the currency has lost its power, this is called devaluation of currency.

For example:
Rate of Exchange
$1 GUY to USD $0.024

If the USD devaluated by 25%, what is the current rate of exchange? So USD has lost 25%, of its buying power, we show this calculation as:

$$100\% - 25\% = 75\% \text{ or } 0.75$$

We say GUY $1 x 0.75 : 0.024USD = GUY 0.75 to 0.024 USD.

This could be said to be the new rate of exchange, where we see that it now takes less of the counterpart currency to purchase the same amount of the currency that has lost its power. However, as mentioned earlier, rates of exchange must be written in a way where either currency is expressed as $1 equivalent of the other currency. We do this by simply writing our ratio as a fraction $0.024 \div 0.75 = 0.032USD$ therefore $1 GUY = 0.032 USD.

Looking at this we can see that GUY $1.00 now has the buying power for more USD than it previously had. (Please take some some time to understand the difference here between revaluation and devaluation). It is basically that when one currency revaluates, the other devaluates and when the other devaluates the counterpart revaluates. It should also be noted how a devaluation is calculated slightly different from the revaluation. We may choose to reconvert our rates of exchange above, so that USD is written as $1.00 to GUY equivalent.
Example:
$1.00 GUY : $0.032 USD

$$\text{GUY\$ } \frac{\overset{31.25}{1.00}}{\text{USD \$0.032}} = \text{Guy \$31.25 : USD \$1.00}$$

1

Practice Activity

Given the Rates of Exchange

€1.00 Euro	:	$1.20 USD
$1.00 CAD	:	$0.75 USD

a). Calculate the value of €500 in USD.

b). Calculate the value of $3,000 US in €.

c). On a vacation in Canada, Johnathan used his credit card to purchase an iPhone valued at CAN $560.00. What is the value of the iPhone in USD?

d). Anna's credit card limit is US $5, 000. If she decides to purchase a smart tv for CAN $2,000 plus 7% tax in Canada using her card, how much money is left on her card for spending?

e). Sarah decided to ship a container of goods to her friend, Lisa, in Europe at a total cost of CAN $1, 200. If Lisa decides to pay her back in Euro, how much money should she send?

f). If the Euro currency undergo a devaluation of 20% against the USD, what would be the new rate of exchange of € : $?

g). If the USD revaluates at 15% against the CAD, how much would CAN $750.00 worth in USD?

BETTER BUY

This is just a little common sense theory included in this book that you may use in your life when it comes to buying and selling.

How do we determine when something is a better buy?

Let us look at the following:

Smiley Orange Juice is sold in cartons of two different sizes at the prices shown below:

Smiley Orange Juice

Carton Sizes	Costs
350ml	$4.20
450ml	$5.13

Explain, showing mathematical working outs, which of the the two cartons are the better buy?

Tips:
- We need to find the cost per each millilitre (cost/millilitre).
- We then compare the cost per each millilitre with both carton sizes to determine the more cost effective buy.

  4.20 / 350ml = 0.012 / ml

  5.13 / 450ml = 0.0114 / ml

From our evaluations above we can satisfy our minds that it is cheaper to buy Carton B than Carton A. Since Carton B is the more cost efficient product, Carton B would be a better buy for our money's worth.

Practice Activity

Smiley Lemonade

Carton Sizes	Costs
350ml	$4.20
450ml	$5.35
500ml	$5.80

Use the table above to answer questions (a) and (b).

a). What is the cost per litre for each of the cartons of lemonade?

b). In your opinion, which of the cartons of lemonade shown above is the most cost-effective buy? (State the reason for your answer)

c). Barbecue potato chips A of 200g, costs $1.76, and another, B, of 900g costs $8.25, which of the two is a better buy?

d). 5 litres of cooking oil costs $2,400. On retail, cooking oil is listed at $520 per litre. Which is best for a family, spending on a tight budget, to choose for a better buy?

e). A man would like to choose between two (2) brands of milk to buy. Brand A holds 750 ml for $4.15 while brand B holds 1 litre for $5.95. Which of the two brands is the better buy?

Answer Sheet

Fractions:

a). $6\dfrac{3}{4}$ b). $1\dfrac{1}{6}$ c). $\dfrac{3}{20}$ d). $4\dfrac{1}{3}$ e). $7\dfrac{1}{2}$ f). $\dfrac{5}{11}$ g). $\dfrac{67}{77}$

Decimals:

a). 3.2   b). 1.17126 x 10^2   c). 0.10   d). 4770   e). (i). 8.909 x 10^-1 (ii). 0.89

f). (i). 6.396021491   (ii). 6.40   g). (i). 210   (ii). 2.1 x 10^2   h). 22.380

i). (i). 1.2   (ii). 1.177x10^0   j).   (i). 0.06   (ii). 0.0566   (iii). 5.6619718 x 10^-2

Percentages:

a). $2,998.80   b). $637.50   c). $1,800   d). 96oz   e). 33.33%

f). (i). 60%   (ii). 270 ripe apples   g). $2,500   h). $40

Interest / Depreciation:

a) (i). $19,200   (ii). $58,560.00   b). $66,951.24   c). $140,100.00   d). $150,663.52

e).$8,238.52

Ratio:

a). 40oz   b). 68 litres   c). 300kg   d). (i). 500:1500:1500   (ii). 1:3:3   e). 250:750:1250

Variations:

a). m = 8   b). b = 24.4   a = 0.39   c). 1.3333 or 4/3   d). 0.5 or 1/2   e). r = 9

f). (i). p = 0.004   (ii). r = 125   (g). x = 5 & y = 40

Foreign Exchange:

a). $600 US   b). € 2,500   (c). $420 US   d). $3,395 US   e). € 750

f). $489.13 US

Better Buy:

a). (i). $0.012/ml   (ii). $0.0119/ml   (iii). $0.0116/ml   b). 500ml is the the most cost effective buy

c). Chips A   d). 5 litres of cooking oil   e). Brand A

Conclusion

Thank you for taking the time-out to engage and study the contents of my publications. It is those who assiduously work hard and spend time knowing their work that usually becomes successful at it. A famous martial artist once said, "I don't fear the man who practices a 1,000 kicks 1 time, but I fear the man who practices 1 kick a 1,000 times". If you keep practicing and study hard you will achieve your goals.

Please take the time out to write me a review on this book at (https://www.dbookship.com). Also subscribe to my mailing list as I will be looking to produce more contents and volumes of this book.

Thanks Much!